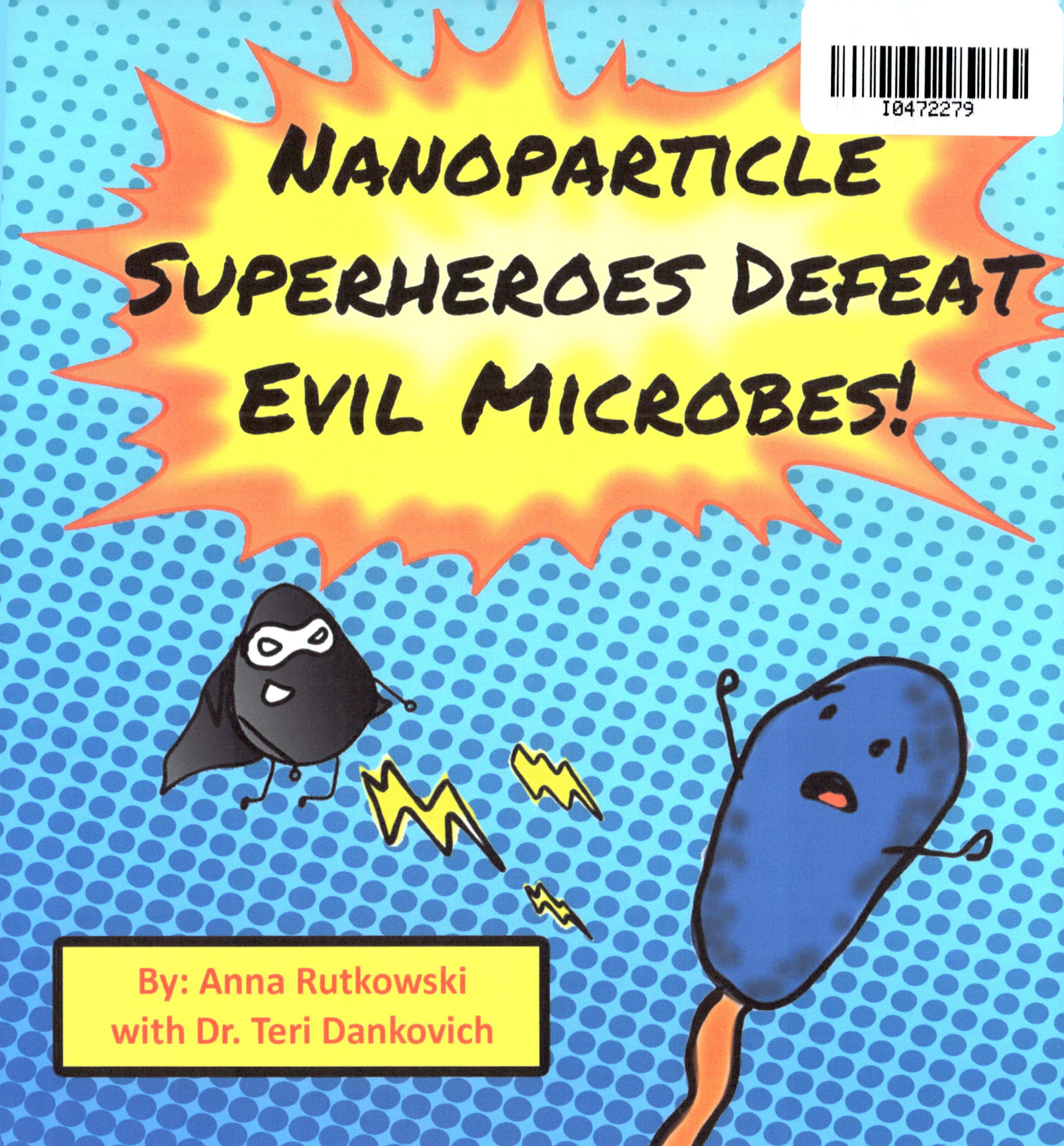

Dedicated to the 663 million people who lack access to safe drinking water.

Text by Anna Rutkowski, Dr. Theresa Dankovich,
 and Dr. Ellen Cavanaugh

Images by Anna Rutkowski

Scanning Electron Microscope image of silver nanoparticles used with permission.

ISBN 978-1-365-85138-4

Grow a Generation
Sewickley, PA 15143
www.growageneration.com

Any and all profits from the sale of this book benefit Folia Water

Dr. Teri in her Lab with a Folia Water Filter

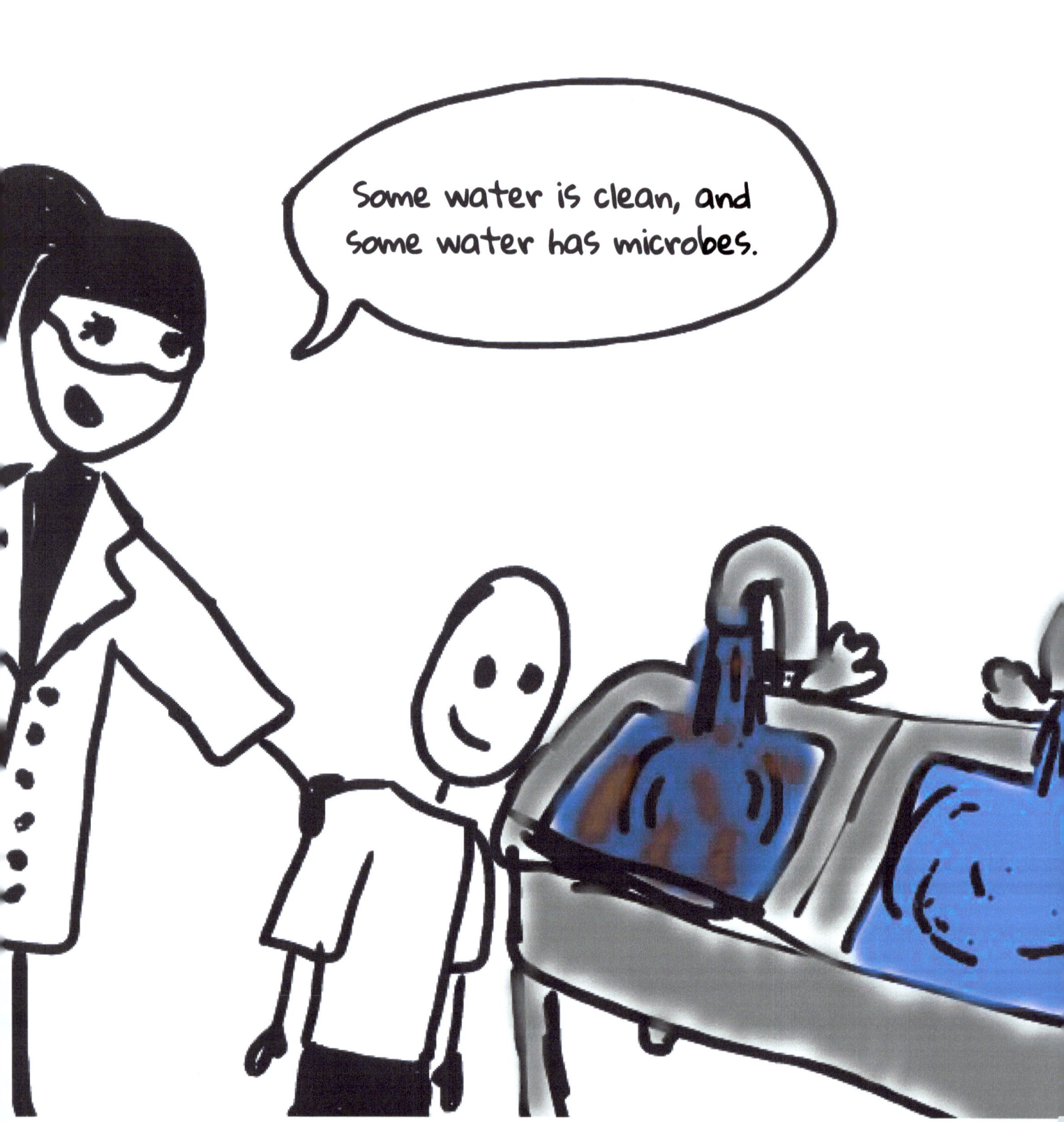

EUREKA! I just made a special paper that destroys microbes and makes the water clean!

The paper has silver nanoparticles!

Silver is a metal AND it is
one of the chemical elements that
make up the whole universe.

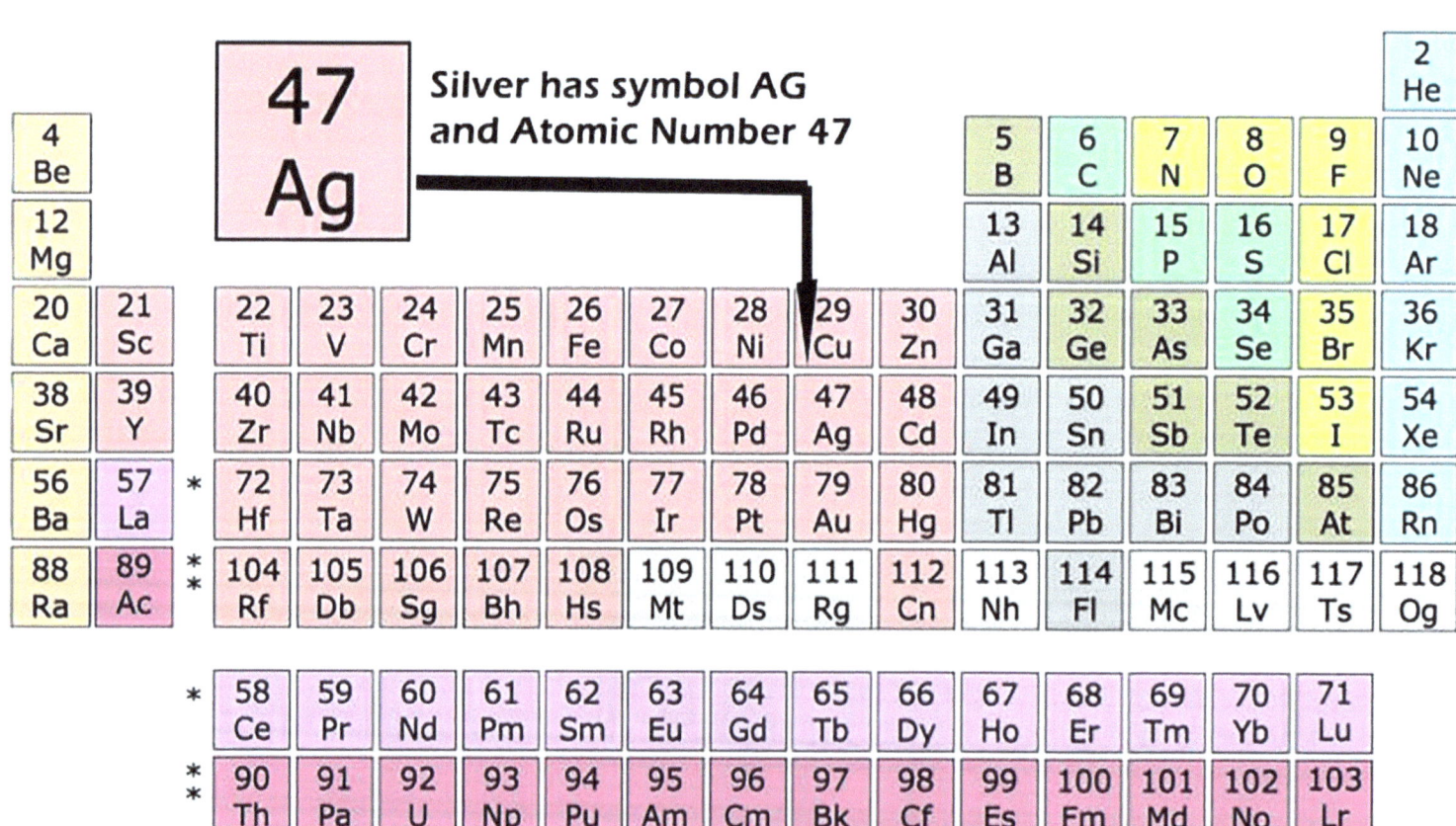

Silver has symbol AG
and Atomic Number 47

Silver NANOPARTICLES are crystalline structures of the element Ag.

SCIENCE TALK

Dr. Teri takes positively charged Ag ions and reduces them to neutrally charged (zero charge) silver atoms, which spontaneously form silver clusters and nanoparticles.

Silver nanoparticles used to be called colloidal silver, and they have been used in many products.

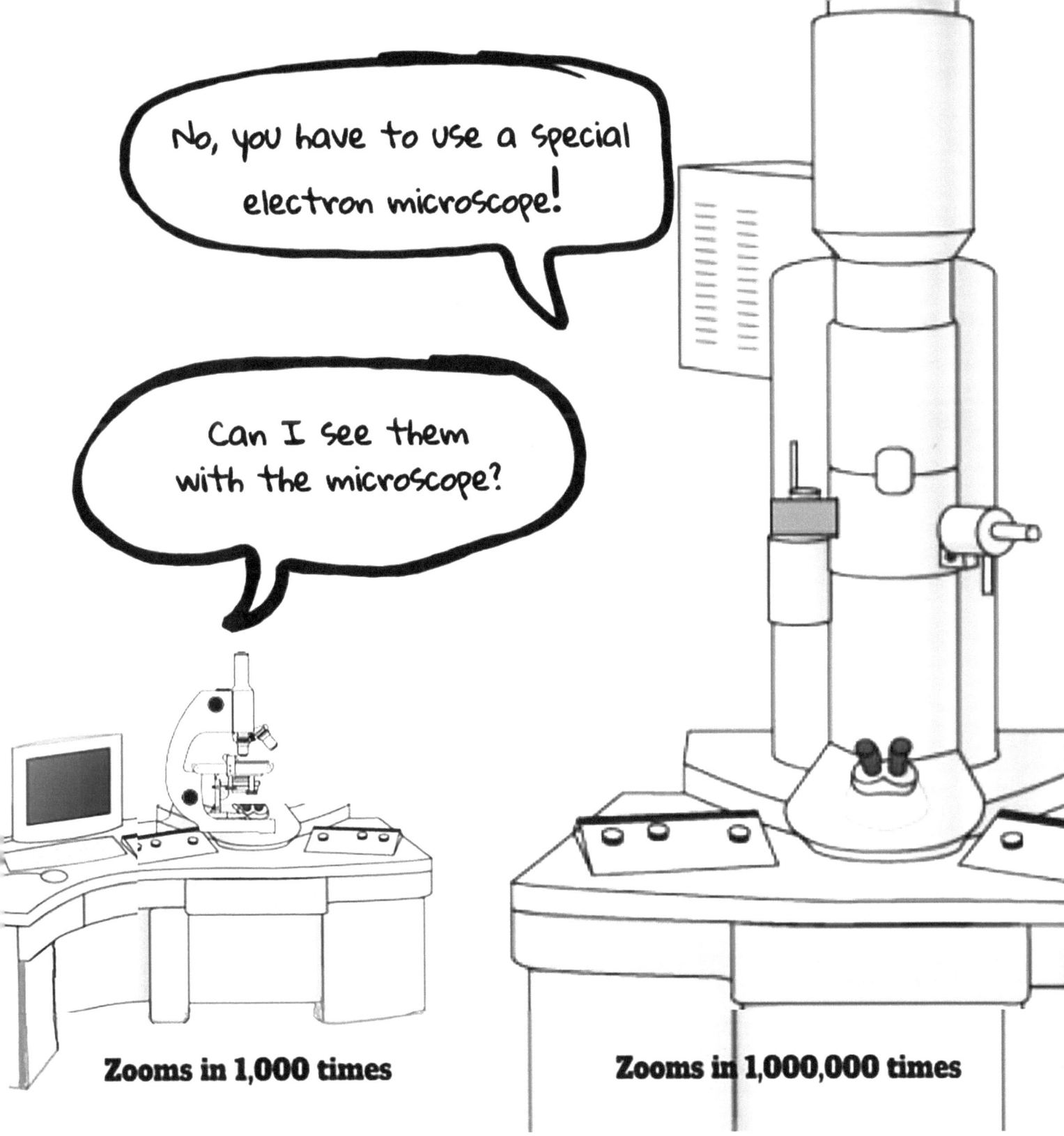

Zooms in 1,000 times

Zooms in 1,000,000 times

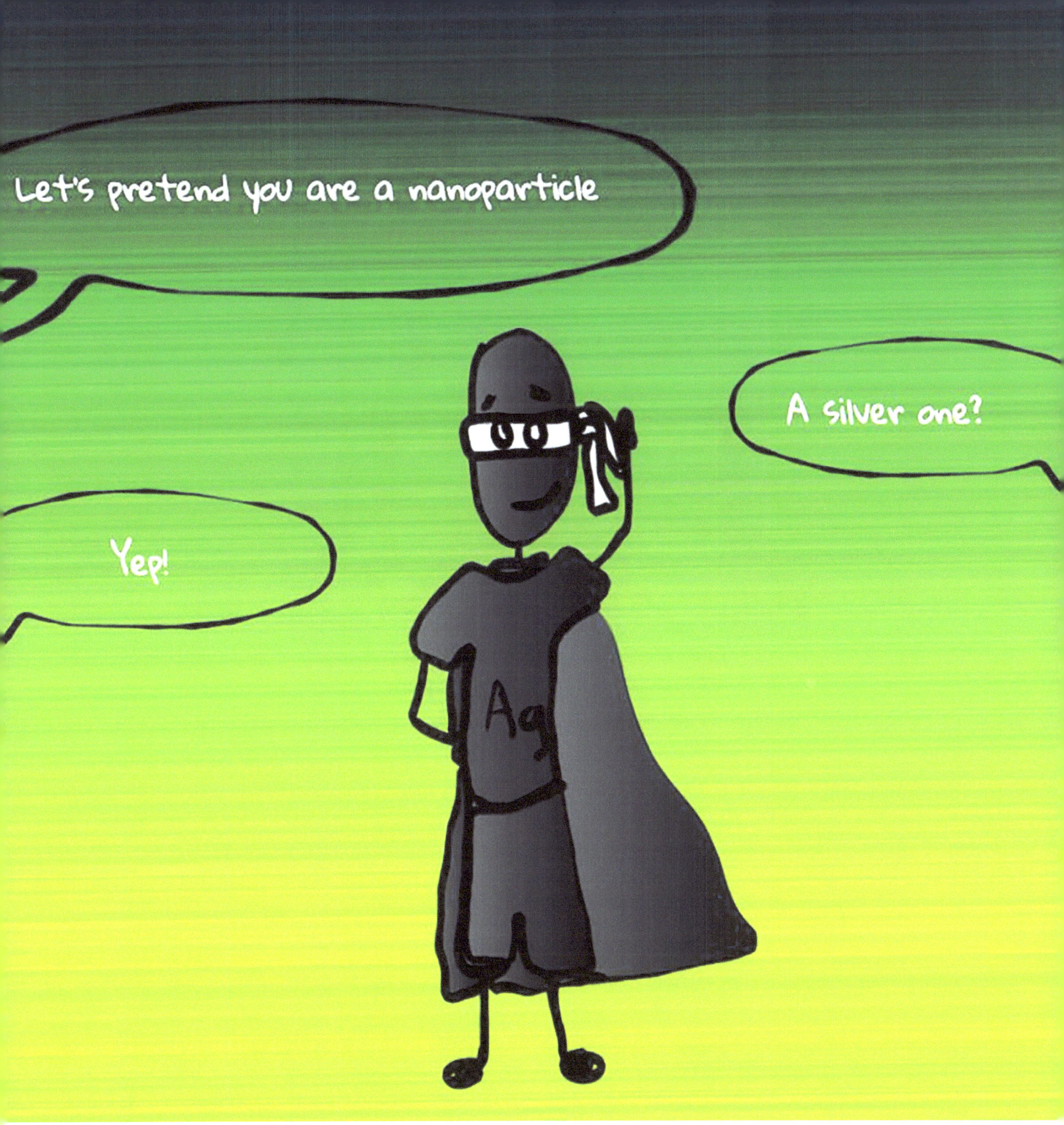
Let's pretend you are a nanoparticle
A silver one?
Yep!
Ag

Let's pretend I make you small.

Are the nanoparticles smaller than the microbes?

Yes! We can see the microbes in a regular light microscope. There are hundreds swimming in a single drop of water.

But that drop of water would be
like an ocean to a silver nanoparticle.

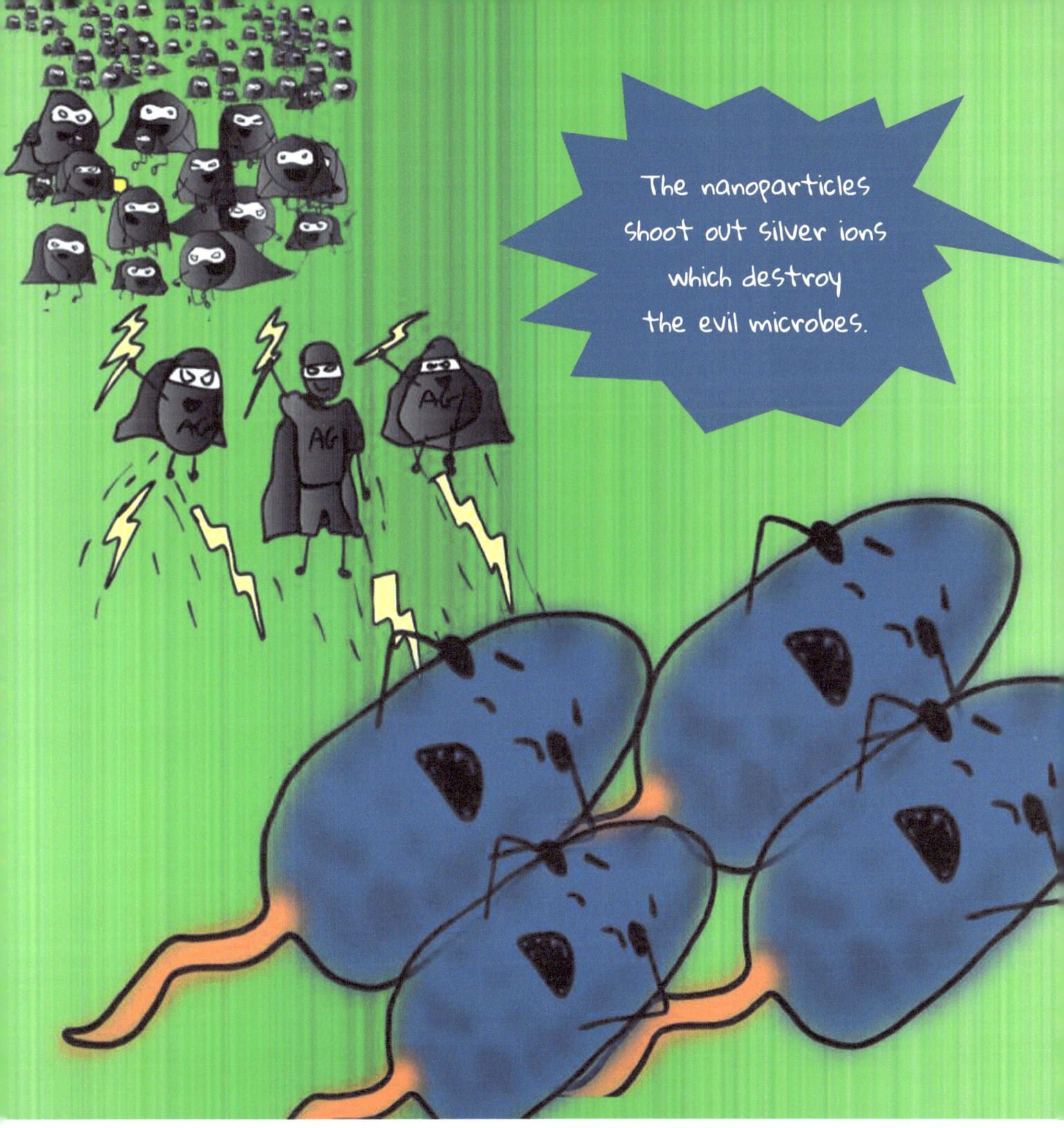

When contaminated water goes into the filter, the evil microbes in the water are destroyed by the silver nanoparticles in the paper.

And safe water comes out the other side.

Your water should be safe if it comes from a treatment plant. Some people get their water from a river.

The microbes and trash in the river make the water unsafe to drink.

- I out of every 6 people living today do not have adequate access to safe water
- I out of every 3 people don't have basic sanitation (for which water is needed)
- Nearly 5,000 children worldwide die each day from diseases that could be prevented with safe water.

You can be a real superhero.
Dr. Teri's Eureka moment has the potential to save lives.
Help us get filters and education to all those families who
do not have clean water.

Profits from this book go to

You can also buy a book of nanoparticle filter paper for a family in need in the form of

Safe Water Books.

Each book contains 26 filters. One book provides a year of safe drinking water for a family and educates them about safe water.

shop.foliawater.com

Dr. Teri (Theresa Dankovich) is the Chief Technology Officer and Chairwoman of Folia Water. She has a B.S. in fiber science from Cornell, an M.S. in agricultural and environmental chemistry from UCDavis, a Ph.D. in chemistry from McGill, and postdocs at the Center for Global Health at University of Virginia and civil and environmental engineering at Carnegie Mellon University. Her doctoral and postdoctoral work has been the invention, laboratory, and field testing of Folia Filters. She created Folia Water to bring her invention out of the lab to those who in need of clean water.

About Grow a Generation

Grow a Generation partners with gifted and talented young people and teachers to make meaningful projects possible.

Faculty, students, and student teams apply in their school to be accepted into the fellowship program.

Once selected, they embark on a year-long odyssey to publish a book, create a digital artifact, or enter a STEM competition.

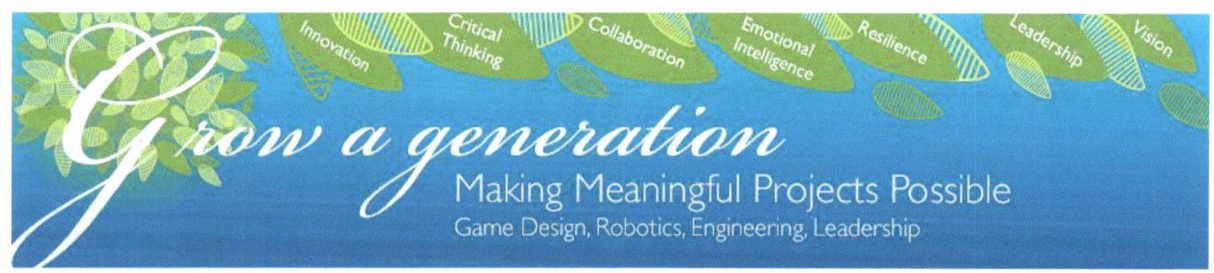

Find out more at growageneration.com.

www.ingramcontent.com/pod-product-compliance
Lightning Source LLC
Chambersburg PA
CBHW051111180526
45172CB00002B/871